Managing Oil Wealth:

The Case of Azerbaijan

Prepared by John Wakeman-Linn, Chonira Aturupane, Stephan
Danninger, Koba Gvenetadze, Niko Hobdari, and Eric Le Borgne

International Monetary Fund
Washington, DC

©2004 International Monetary Fund

Production: IMF Multimedia Services Division
Cover Design: Luisa Menjivar
Typesetting: Alicia Etchebarne-Bourdin
Cover Photo: AFP Photo/Anadolu Ajansi/Riza Ozel

Cataloging-in-Publication Data

Managing oil wealth : the case of Azerbaijan / prepared by John Wakeman-Linn . . . [et al.].
   — [Washington, D.C. : International Monetary Fund, 2004]
            p.   cm.

            ISBN 1-58906-308-2
            Includes bibliographical references.

   1. Petroleum industry and trade—Azerbaijan.   2. Natural resources—Azerbaijan—
Management.   3. Natural resources.   I. Wakeman-Linn, John.

HD9576.A982 M26 2004

Price: $20.00

Please send orders to:
International Monetary Fund, Publication Services
700 19th Street, NW, Washington, DC 20431, U.S.A.
Telephone: (202) 623-7430   Telefax: (202) 623-7201
Internet: http://www.imf.org

Contents

Boxes

Figures

Tables

Acknowledgments

This paper has benefited from comments by Julian Berengaut, Oleh Havrylyshyn, Juha Kähkönen, Rolando Ossowski, Mark Flanagan, Michael Mered, Basil Zavoico, Christian Petersen, and Peter Thomson. The authors would like to thank Malina Savova for research assistance; Grace Moss for assistance in putting the manuscript together; and Maria Cecilia Pineda for document preparation.

The authors would also like to thank the Azerbaijan International Operating Company (AIOC) for providing us with valuable data. Sean Culhane perfomed an editorial review and coordinated the production of this volume.

1 Introduction

Azerbaijan has a substantial endowment of oil and gas deposits, estimated to be the third largest in the Caspian region. Oil production in Azerbaijan is projected to increase sharply starting in 2005, and to reach a peak in 2009 of 1.3 million barrels per day, or four times current production. Gas production is expected to increase in 2006 following the development of the Shah Deniz gas field and construction of the related gas pipeline, reaching an annual peak of 20 billion cubic meters (bcm) in 2010. Even under conservative assumptions about international oil and gas prices, the expected revenue windfall to the government of Azerbaijan over the next 20 years is substantial. However, given the known reserves and production profile, oil and gas revenues are expected to peak at the turn of the decade and decline gradually thereafter, and to return to current levels by 2024.

Azerbaijan faces the challenging task of reducing its dependence on short-lived and potentially volatile oil revenue. It is vital to the country's economic future that the government manage this revenue in a way that allows the diversification of the economy, in order to ensure a steady increase in the living standards of the Azerbaijani population. This is essential not only because of the temporary nature of the boom, but also because the oil sector—while a substantial source of revenue for the country—is not a source of much employment, with only 1.1 percent of the Azerbaijani labor force employed in the sector in 2001.

Few countries that have been heavily dependent on the oil sector have succeeded in managing oil wealth in a manner that allowed the simultaneous development of the non-oil sector. Norway and Indonesia are frequently cited as exceptions. The Norwegian economy has experienced solid economic growth for the last three decades. The fact that Norway was already a developed and diversified industrial economy, with a long tradition of democracy, a market-oriented economy, significant and varied nonenergy exports, and solid and mature institutions may largely explain its success. Indonesia also had other export commodities (rubber, coffee, timber) and sought to ensure that they continued to generate considerable income. Indonesian authorities undertook prudent macroeconomic policies, which at times required significant expenditure cuts and correction of misaligned exchange rates in order to adjust to volatility in oil revenues (Appendix 1).

The list of countries that failed to avoid the problems associated with natural resource booms is long, including Nigeria, Angola, Algeria, Mexico, Venezuela, and Ecuador. For most countries, natural resource booms were the impetus for economic disorder and crises (some examples are discussed in Appendices 2 and 3). It is crucial that Azerbaijan design and adopt prudent and coordinated macroeconomic policies and institutional reforms that take into consideration the experience of these countries in order to avoid the mismanagement of natural resource wealth and its implications.

This paper aims to provide a guide to the management of Azerbaijan's expected natural-resource-generated windfall, based largely on the lessons that can be drawn from the experiences—mostly negative—of other countries. Chapter 2 discusses the economic theory of natural resource booms and briefly explains the standard Dutch disease phenomenon. Chapter 3 discusses common characteristics of policies leading to the mismanagement of natural resource wealth in natural-resource-abundant countries. Chapter 4 explains the institutional arrangements of oil revenue management in Azerbaijan and estimates oil and gas revenue prospects for the country. Chapter 5 outlines a medium- and long-term strategy for oil wealth management in Azerbaijan, building on the lessons in Chapter 3. Chapter 6 concludes.

2 Economic Theory and Natural Resource Booms

Studies of experiences of countries rich in exhaustible natural resources reveal that natural-resource-driven booms have often led to deterioration in macroeconomic performance and uneven development of industry. Sachs and Warner (1995) provide empirical evidence that economies with abundant natural resources have tended to grow less rapidly than economies with scarce natural resources. Large foreign exchange inflows due to the exploitation of natural resources often turn into a curse for the country, particularly if they are mismanaged. This adverse effect of natural resources has been called the "Paradox of Plenty" (Karl, 1999).

This "natural resource curse" is the result of the interaction of two factors. The first is policy mismanagement, often a direct result of the easily available revenues that both lead to rent-seeking behavior and reduce pressures for necessary economic reforms. The second factor is the "Dutch disease," which leads to a decline in other tradable sectors. The Dutch disease phenomenon refers to the loss of competitiveness, or deindustrialization, of a nation's economy that occurs when a natural-resource-inspired boom raises the value of the domestic currency, making manufactured goods less competitive, increasing imports, and decreasing exports. These factors interact, as the Dutch disease makes effective economic policy management both more difficult and more essential.

A simple two-industry model can describe the Dutch disease phenomenon. Suppose two industries are producing goods traded at prices determined in the international market. The industries employ labor from a common pool, combined with a factor specific to each sector and in fixed supply. If the world price of the output for one of these industries rises, the returns to that industry will increase, pushing up wages in that industry. The marginal productivity of labor in the booming industry will increase and attract labor away from the nonbooming industry. This change in the sectoral composition of labor is called the *resource movement effect* of the boom (Corden, 1992). Higher wages in the booming industry will also squeeze profits of the other traded-goods industry that has not experienced a rise in price. As a result, the production of the second industry will decline.

The two-sector Dutch disease model can be extended to three sectors to more accurately reflect the real world: a traditional traded-goods industry, a booming traded-goods industry, and a nontraded-goods industry. Higher real incomes from the booming sector lead to increased expenditures on both traded and nontraded goods. This does not cause the price of traditional traded goods to rise, as their price is determined in the international market. By contrast, the price of nontraded goods is set in the domestic market and does rise due to increased demand. This is called the *spending effect* of the boom (Corden, 1992). This *real appreciation* (defined as an increase in the real exchange rate, the price of nontradables relative to tradables) leads to a *resource movement* from the traditional traded to the nontraded sector, an expansion in the nontraded-goods industry, and a contraction in the traditional traded-goods industry—or Dutch disease.

This real exchange rate (RER) appreciation is almost inevitable during a boom, and is required to maintain money market equilibrium. Saving some of the income from the booming sector abroad in the form of foreign assets, or using it to pay off external debts ahead of schedule, could, however, curb domestic spending, thereby limiting the RER appreciation and its adverse consequences.

As noted above, the pressures of Dutch disease make effective policy management—in particular, the restructuring of traditional and other trade sectors, and the creation of a favorable environment for the development of other industries—even more important, as the government must seek to offset the effects of the RER appreciation on the competitiveness of other industries. At the same time, the easy availability of these revenues reduces political pressure for such reforms, as the revenues can be used to finance the status quo. Indeed, Sala-i-Martin and Subramanian (2003) argue that the main channel through which oil wealth leads to the Paradox of Plenty is through its adverse effect on the development of institutions.

The importance of prudent management of revenues accruing from the booming sector and the avoidance of Dutch disease is magnified in circumstances where natural resource revenues are expected to be short-lived, as is the case in Azerbaijan. If the boom leaves behind a noncompetitive and contracted traded-goods industry, export incomes will not be able to finance an expanded public sector and the country's need for foreign exchange in the period following the natural resource boom. This will make an economically painful and politically difficult adjustment unavoidable.

3 Country Experiences with Managing Natural Resource Windfalls

It is ironic that what should be a blessing has often turned into a curse. A natural resource boom, effectively managed, should provide a country with the resources to finance essential economic reforms, including a cushion for the vulnerable against the impact of those reforms. Unfortunately, the experiences of natural-resource-abundant countries show that this has not typically been the case. In general, rather than using natural resource revenue to finance the development of other aspects of the economy, authorities have usually tended to act based on optimistic assumptions about the size and extent of natural resource booms, and to use these optimistically forecast revenues to finance consumption.

Decisions regarding the spending of natural resource revenues should be based on the likely duration of the resource boom, the expected income (subject to price assumptions), extraction costs, and the time horizon during which exhaustible resources may be depleted. In light of the uncertainties associated with these estimates and the unpredictable path of the terms of trade, it would be logical to take a cautious stand and forgo present consumption in favor of security against unfavorable developments in the future. However, the experience of resource boom countries shows that, generally, authorities tend to act on optimistic assumptions. Below are examples of policies that have been common in both developed and developing countries managing windfalls from natural resource booms during the 1970s and 1980s, and the implications of these policies.

Authorities frequently did not utilize higher natural resource revenues to reduce budget deficits, and tended to spend them inefficiently. Counting on high current and future income, expenditures were brought into line with this anticipated high-income level within a relatively short period of time. As a result, the budget deficit widened (Mexico, Nigeria). In some cases, countries borrowed heavily against their anticipated future oil income (Algeria, Venezuela). In addition, authorities often granted large wage increases to public sector employees (Trinidad and Tobago, Nigeria, Venezuela) and created new government structures with new positions. Later, financing increased wage bills contributed to higher inflation, as the authorities found it difficult to reverse nonsustainable expenditure levels once the windfall subsided.

In expectation of continued revenue from the resource boom, authorities undertook ambitious public domestic as well as foreign investment projects with low economic rates of return, politically attractive payoffs, inadequate screening, and undiversified risk (Algeria, Trinidad and Tobago, Nigeria, Iran, Côte d'Ivoire). Often such projects served the interests of well-connected individuals. Furthermore, the maintenance costs of these large, nonviable projects were underestimated, and following the resource boom, the government faced the difficult trade-off of sharply reducing other expenditures, postponing their implementation, or stopping project maintenance completely (Nigeria, Mexico, Indonesia). The discontinuation of such projects would leave valuable financial resources wasted and former employees jobless.

The windfall associated with the natural resource boom weakened the authorities' commitment to undertake necessary restructuring of underdeveloped sectors. Subsidies to these sectors, which were easy to finance during the boom, became hard to maintain after revenues from the booming industry declined. The ailing sectors would have functioned without subsidies, or at least with substantially smaller subsidies, had they undergone the necessary restructuring during the boom times. In general, authorities of countries endowed with rich natural resources tended to be overly confident and underestimated the need for the creation and development of growth-conducive institutions and infrastructure (Gylfason, 2001).

The exploitation of natural resources often promoted rent-seeking behavior, especially under conditions of inappropriately defined property rights and lax law enforcement. Windfall revenue from an export boom also contributed to social problems such as corruption and caused further imbalance in the income distribution. The neglect of the environmental impact of natural resource exploitation led to unrecoverable damages, requiring a high cost of restoration (Nigeria, Ecuador, Indonesia).

Following a natural resource boom, stop-gap policies adopted to counteract the resultant economic imbalances tended to have a further negative impact on the economy. After an adverse terms-of-trade movement, the traditional traded-goods sector was not in a position to earn the necessary foreign exchange, and authorities employed protectionist policies such as restrictive quantitative controls, import quotas, higher tariffs, and bureaucratic barriers to prevent foreign exchange outflows. Such inward-looking policies hurt the manufacturing sector and made repayment of external debt difficult (Ecuador, Nigeria, Mexico).

Many natural-resource-rich countries created savings or stabilization funds with the aim of protecting the domestic economy from a volatile path of natural resource revenues or for saving the windfall resources for future generations.

One study of the experience of such funds in five selected countries (Fasano-Filho, 2000) showed that saving natural resource revenues in such funds and investing the funds' resources abroad might have contributed to limiting domestic spending pressures (*spending effect*) and reducing real exchange rate appreciation during periods of a rising price for the natural resource (Norway, Chile). The same study concludes that the experience with stabilization funds has at times been less positive, due to frequent changes in the funds' rules and deviations from their intended purposes (Venezuela, Oman). Success did not lie in the creation of such funds, but rather in fiscal discipline and sound macroeconomic management.

To avoid the consequences of a mismanaged natural resource boom, Azerbaijan will need to make important decisions about consumption, savings, and investment policy and not relax its attention to underlying structural problems. If the country does not prepare itself properly before the boom occurs, this may at the end bring economic disorder.

4 The Oil Sector in Azerbaijan

Azerbaijan has a rich natural resource endowment and a long history of oil and gas exploration. Oil and gas reserves in the country are estimated to be the third largest in the Caspian region.[1] Oil production peaked in 1941 at 172 million barrels of oil, or almost 75 percent of the output of the Soviet Union. Production then declined steadily, dropping off sharply in the final years of the Soviet Union. Only in the late 1990s did discoveries of new oil and gas reserves lead to a turnaround in output, driven primarily by foreign investment from international partners.

The management of the oil sector falls into two categories. Soviet-era oil and gas fields are operated by the state oil company (SOCAR) with weak prospects for a further expansion of production. Current production levels of the Soviet-era oil and gas fields are at around 177,000 barrels of oil per day (bpd) and 4.4 billion cubic meter of gas. Most new oil and gas fields are developed and managed under the leadership of international partners. Income from these operations is shared with the government according to predetermined production-sharing agreements (PSAs).

Azerbaijan has signed a number of PSAs for the exploration and development of the country's hydrocarbon resources. In 1994, the government signed its first foreign-partnered PSA, popularly referred to as the "Contract of the Century," with an international consortium, the Azerbaijan International Operating Company (AIOC), to develop the Azeri-Chirag-Guneshli (ACG) oil fields in the Azerbaijan sector of the Caspian Sea. In addition, 21 other PSAs have been signed and ratified since then for the exploration and development of the country's onshore and offshore hydrocarbon reserves.

---

[1]Source: U.S. Department of Energy (as of July 2002).

Despite some significant oil and gas discoveries, most PSAs have yet to find commercially viable oil or gas deposits. In 1999, potential recoverable natural gas resources in excess of 14 trillion cubic feet were confirmed in the Shah Deniz field, reportedly the largest natural gas discovery since 1978 for British Petroleum.[2] In 2002, total oil reserves in the ACG fields were determined to be higher than previously anticipated, at 5.4 billion barrels. Current production stands at around 130,000 bpd, with peak production of slightly over a million bpd anticipated at the turn of the decade from the ACG oil fields. The recently initiated pipeline project from Baku to Ceyhan has greatly enhanced these prospects. However, the success of the other 20 PSAs has been limited. A few PSAs have been abandoned due to the lack of commercially viable oil deposits. To date, only one other PSA (Salyan Oil Consortium) is in the production stage, with 8.2 thousand tons of oil exports in the first quarter of 2003, while some others are under discussion for abandonment.

A. Current Institutional Arrangements for Managing Oil Revenues

The separate operational structures for old and new fields have led to a division in the management of oil and gas revenue. Figure 1 summarizes the government's main oil revenue sources and the two government bodies—the state budget and the State Oil Fund—involved in the management of oil and gas revenues. The consolidated government receives *profit oil* and *income tax* from the development of new fields, as spelled out in the PSAs with international partners. The profit oil component of these flows accrues to the State Oil Fund while the income tax (personal and profit tax) component flows to the state budget.[3] The old fields, operated by SOCAR, generate *income tax* revenue, which is paid to the state budget.

The State Oil Fund (SOFAZ) is the key institution for the management of oil wealth in Azerbaijan (Box 1). It was established in 1999 as an extrabudgetary fund in order to ensure transparency in the management of oil revenue and to curtail the use of assets. Its main purpose is to save funds for future generations, but assets are also used for investment projects. As of end-March 2003, SOFAZ's total assets amounted to US$727 million.

---

[2]Source: Azerbaijan International Operating Company (AIOC). Also see http://www.dnv.com/publications/oilgas_news/by_subject/General/ShahDeniz.asp.

[3]Personal income tax accrues according to the tax code while profit tax obligations are defined in the PSA.

Figure 1. Sources of Oil and Gas Revenue in Azerbaijan

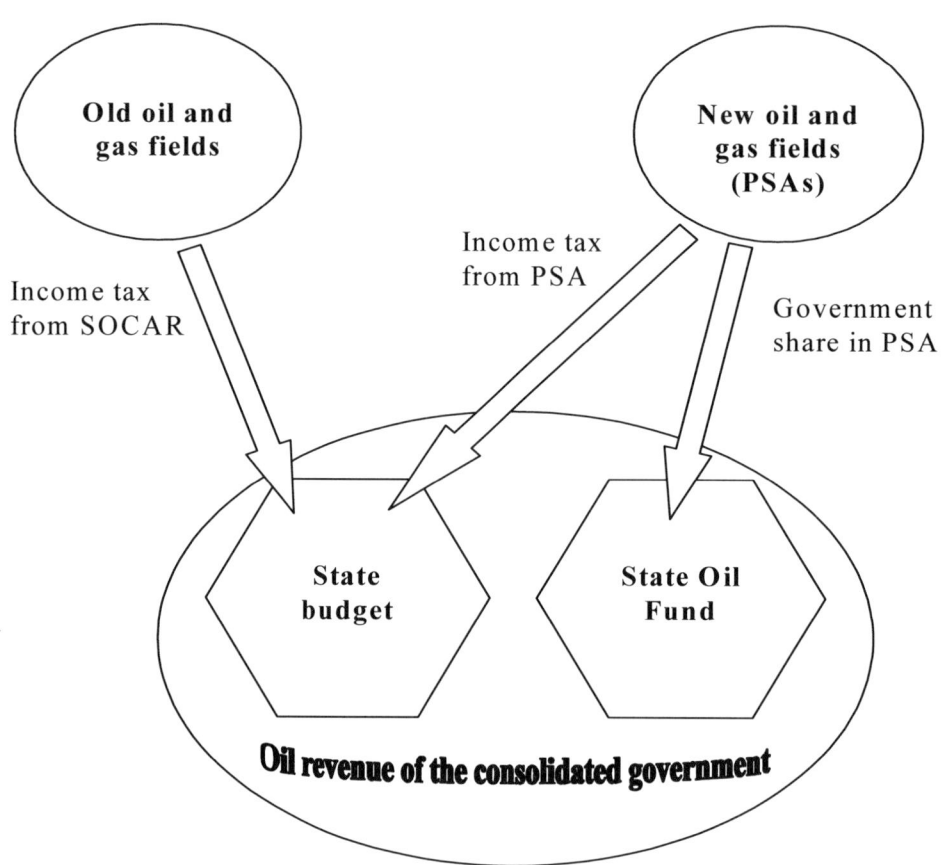

Source: Azerbaijani authorities and IMF staff.

Significant additional oil revenue accrues to the state budget primarily from SOCAR tax payments. In 2002, oil- and gas-related revenues of the state budget were US$340 million, about US$100 million higher than receipts of SOFAZ. However, as SOCAR's production declines over time and new fields are developed, inflows to the oil fund will dwarf state budget revenue as early as 2006 (see discussion below). Unifying the government's management functions for oil revenue should be an important consideration in view of the challenges arising from the expected oil boom, as discussed below.

Box 1. State Oil Fund of the Azerbaijan Republic

The State Oil Fund of the Republic of Azerbaijan (SOFAZ) was established in 1999 as an extrabudgetary institution. Its main objective is the professional management of oil- and gas-related revenues for the benefit of the country and its future generations—i.e., savings. The inflow and outflow rules of Azerbaijan's oil fund have been designed to reflect this feature and to save a large part of government oil and gas revenue. SOFAZ receives all government revenues associated with the post-Soviet oil and gas production fields. The oil fund has no immediate stabilization objective and net flows are not related to the oil price level or a budgetary position. On the outflow side, Azerbaijan's oil fund rules currently prohibit spending in excess of inflows in any given year. A conservative expenditure policy has ensured a steady growth of savings in the fund. Asset management regulations require that financial assets be kept offshore at highly rated banks. The fund is not permitted to extend credits to private or state organizations and assets cannot be used as a guarantee against any obligation.

In order to reduce political pressures to spend windfall oil revenues rapidly, the government established the oil fund under direct presidential control. The members of SOFAZ's supervisory board are appointed by the President of Azerbaijan. An independent auditor conducts an annual audit of the fund, and the audit report is made public. SOFAZ reports quarterly in the press on total inflows received, expenditures, and interest earned. To strengthen the legislative foundation of SOFAZ, its budget and asset management rules were approved by Parliament in June 2003 as amendments to the Budget System Law. The creation of an oil fund in Azerbaijan has had a positive impact on fiscal discipline and contributed to better transparency and accountability of oil revenue management.

B. Prospects for Government Oil and Gas Revenues

Substantial, but short-lived, revenues associated with the development of the oil and gas fields are expected to accrue to the country from (1) profit oil and profit gas according to the terms of the ACG and Shah Deniz PSAs, (2) profit tax payments from partners under the PSAs, and (3) SOCAR tax payments. Figure 2 presents the accumulation of natural resource revenue from the three sources for different production scenarios. Projections for SOCAR revenue are based on a slightly declining output profile consistent with current production expectations. Profit tax and SOFAZ inflows of profit oil and profit gas associated with the development of the ACG and Shah Deniz oil and gas fields are calculated under three oil production profiles consistent with proven, probable, and possible oil reserve[4] estimates for the period 2000–25.

---

[4]Proven, probable, and possible reserves are defined as oil deposits that are considered 90, 50, and 10 percent likely, respectively.

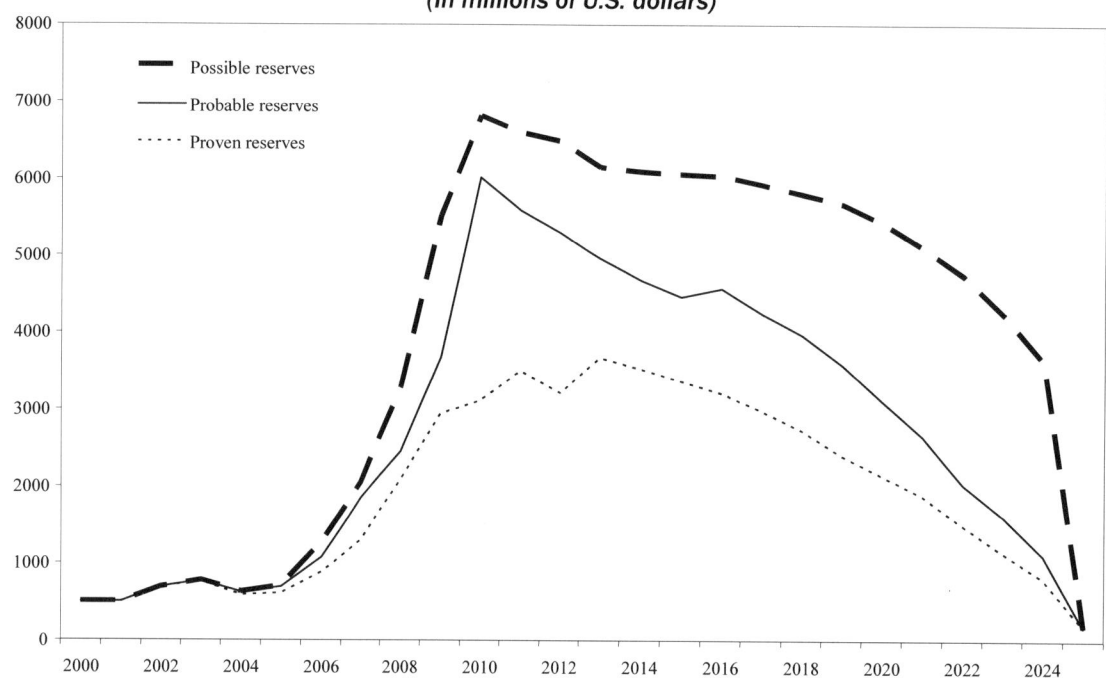

Figure 2. Azerbaijan Oil and Gas Revenues, 2000–25[1]
(In millions of U.S. dollars)

Sources: Azerbaijan International Operating Company and IMF staff estimates.
[1]Based on June 2003 World Economic Outlook oil and gas price assumptions and excluding asset management revenue.

The profile of Shah Deniz gas production is the same for all three oil production profiles and is consistent with the Shah Deniz Stage 1 production profile and associated sales agreements for the time period.[5] All three scenarios use World Economic Outlook (WEO) oil and gas price assumptions as of end-June 2003. Under the conservative proven reserves scenario, substantial oil- and gas-related revenues are expected to accrue, with revenues increasing more than sevenfold during the period 2000–13. However, this sizable increase in revenues is short-lived, as following the peak in 2013, inflows to SOFAZ decline fairly rapidly. ACG PSA-related oil revenues are expected to end after 2024, following the depletion of the ACG oil reserves, absent a significant new hydrocarbon discovery.

[5]Gas-related revenues over the period 2000–24 are projected to commence in 2006 with the first gas exports from the Shah Deniz field, but are negligible in comparison to the revenues accruing to the government from the ACG PSA.

Figure 3. Azerbaijan Oil and Gas Revenues Under Conservative Price Assumptions, 2000–25[1]

(In millions of U.S. dollars)

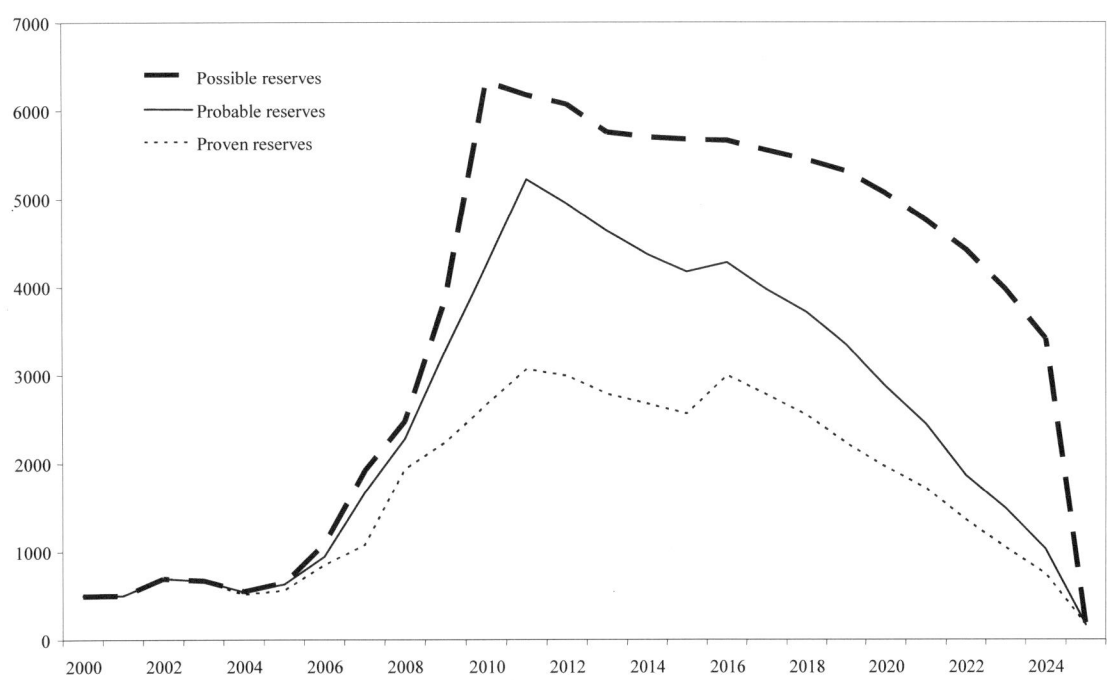

Sources: Azerbaijan International Operating Company and IMF staff estimates.
[1]Based on a fixed US$20/barrel oil price assumption, gas price assumptions more conservative than World Economic Outlook and excluding asset management revenue.

Even under more conservative price assumptions the expected revenue stream is large. Figure 3 presents the same three production scenarios, but assumes a fixed US$20 per barrel oil price and gas price assumptions more conservative than WEO's. Under these more conservative assumptions, oil- and gas-related revenues are still expected to increase almost sixfold during the period 2000–13 in the proven reserves volume scenario. As these charts indicate, even under a wide range of production profiles and different oil and gas price assumptions, a similar pattern of oil and gas revenue receipts emerges: an accrual of substantial revenues during a relatively short period of time.

5 Strategy for Managing Oil Wealth

With this significant asset stock, the key challenge for the government will be to strike the right balance between current expenditures and conserving assets for future generations. The government's decision about how much to spend and how much to save involves important trade-offs. For example, addressing poverty and infrastructure needs quickly may alleviate poverty in the short run, but may pose a risk to macroeconomic stability and damage the long-term growth potential of the non-oil sector. On the other hand, using a measured approach to the use of oil revenue assets will require a strong political commitment and public engagement to fend off political pressures for increased spending, and could be hard to justify in the face of undeniable and substantial needs.

One way to address this challenge is to separate the expenditure problem into a long-term strategy focused on conserving financial assets for future generations, and a medium-term strategy *within* this framework aimed at addressing immediate policy objectives. The long-term policy would provide an expenditure envelope for the medium term based on purely long-term considerations, while the medium-term plan would be concerned primarily with macrofiscal stability and management. A viable expenditure strategy would then require that both medium- and long-term considerations are met. Determining expenditure priorities and fine-tuning medium- and long-term policies will be an ongoing problem for the government.

Separating this problem into a two-step process reduces its complexity and makes it more tractable. The proposed approach aims to provide a framework for this task and guidelines on the consolidated government deficit, but it does not deliver a specific expenditure plan, as that will require critical political decisions. As economic conditions change, these plans will have to be regularly revisited and modified.

The next section discusses the determination of a long-term strategy that sets a ceiling for feasible expenditure plans in the medium term. The subsequent section discusses medium-term policy options.

A. Long-Term Consumption and the Sustainable Non-Oil Balance

Any long-term savings objective limits the amount of available assets for immediate consumption in order to spread out their use over time. The concept of a sustainable non-oil deficit ceiling translates this restriction into an upper bound for the permissible government deficit consistent with the savings objective. In other words, it defines what the government can afford to spend over the *long term* without exhausting its assets, and corresponds to a path of expenditures that can be financed from the use of oil revenue. The non-oil balance—as a direct measure of the level of activity that is financed from oil revenue—is a crucial guide for fiscal policy in oil-producing countries. Other common measures of government activity, such as the overall balance or the current balance, obscure the actual fiscal stance as they are affected by changes in oil prices (Barnett and Ossowski, 2002).

Various economic theories exist on the optimal distribution of oil wealth consumption over time, sometimes providing conflicting advice. The conclusions of these theories can, however, be classified in two categories: those advocating a front-loaded spending approach and those advocating a back-loaded approach. Arguments in favor of the former category are (1) the need to "jump-start" an economy, which is in a development trap;[6] (2) political economy factors, such as the extent of initial income inequality;[7] and (3) intergenerational equity considerations (assuming public expenditures are sufficiently productive to increase future economic growth). Theories supporting back-loaded expenditure profiles emphasize capacity constraints on the management of rapid expenditure increases, the Dutch disease problem, governance issues, intergenerational equity (assuming public expenditures out of the fixed oil wealth do not generate positive returns, e.g., because of corruption or waste or because they are focused on current consumption), and unknown future liabilities.[8]

Fiscal rules can then be applied to suit the specific characteristics of a country in terms of the optimal distribution of oil wealth over time. Two simple rules are explored to illustrate their implications. One rule is to guarantee a constant real expenditure stream with a slightly more back-loaded variation of constant real

---

[6]That is, the need for a critical mass (e.g., of human and physical capital) before economic "takeoff" can occur (see Azariadis and Drazen, 1990).

[7]Theoretical models have shown how income inequality can exert a negative effect on investment and on subsequent economic growth because it provides strong incentives for redistribution policies, which hurt growth-promoting investment (e.g., Persson and Tabellini, 1994). Empirical evidence has confirmed such a negative relationship, at least for democratic regimes.

[8]For instance, pension liabilities, whose future cost is subject to a large degree of uncertainty (political considerations could significantly affect the payout to future pensioners).

per capita expenditures.[9] This strategy has the advantage of being simple and intuitive. A second, more sophisticated, fiscal rule aims to generate a permanent income stream by preserving a stock of wealth. The rule postulates that total oil wealth—the sum of the (underground) oil wealth and financial wealth—has to remain constant in real terms over time (or in real per capita terms in a second version of this rule); as oil resources are exploited and the value of oil in the ground declines, a fraction of the oil wealth needs to be saved and turned into financial wealth so as to keep total oil wealth constant. The advantage of this rule is that it captures the full value of oil wealth—namely oil in the ground and financial wealth—and it enables society to permanently benefit from its use. Its sustainable deficit path is however determined less intuitively and limits its use for public discussion.

Figures 4 and 5 provide estimates of the sustainable non-oil deficit for the two constant expenditure fiscal rules (in real and per capita terms) and for two oil price scenarios. The three lines within each graph correspond to the different production paths. Details on the derivations of the sustainable deficits paths can be found in Appendix 4. The projections assume a long-term real interest rate of 5 percent, a long-term inflation rate of 2 percent, a non-oil sector growth rate of 5 percent, and population growth at 1 percent per year.

There are remarkably few differences between the results for the expenditure rule and the wealth conservation strategy. In virtually all scenarios the sustainable non-oil deficit ceiling is high in the medium term, allowing a non-oil deficit of around 20 percent of GDP on average in 2004–07. The variation around this ceiling is, however, fairly large and ranges from 10 to 30 percent of GDP. This emphasizes the need for careful monitoring and regular updates of the sustainable deficit ceiling to provide frequent feedback for medium-term expenditure plans.

The front-loaded expenditure pattern in all scenarios reflects the fact that in early years the government can afford high expenditures in anticipation of large oil revenue receipts. The simulations however also show a marked drop-off in the sustainable ceiling over time. Two factors drive this result. First, as oil revenue materializes, long-term savings objectives require an accumulation of assets to permanently finance the desired expenditure stream or to maintain a given wealth stock. Second, the decline of the deficit as a share of GDP is also affected by the rapid growth of the oil sector and thus GDP. In real terms the deficit ceiling increases.

---

[9]Simplicity and transparency are desirable features for a rule, making it easier for a government to explain to the public and easier for the public to monitor adherence to the rule, thereby enhancing its credibility.

Figure 4. Azerbaijan: Sustainable Non-Oil Deficit Ceiling—Constant Expenditures, 2004–24
(In percent of GDP)

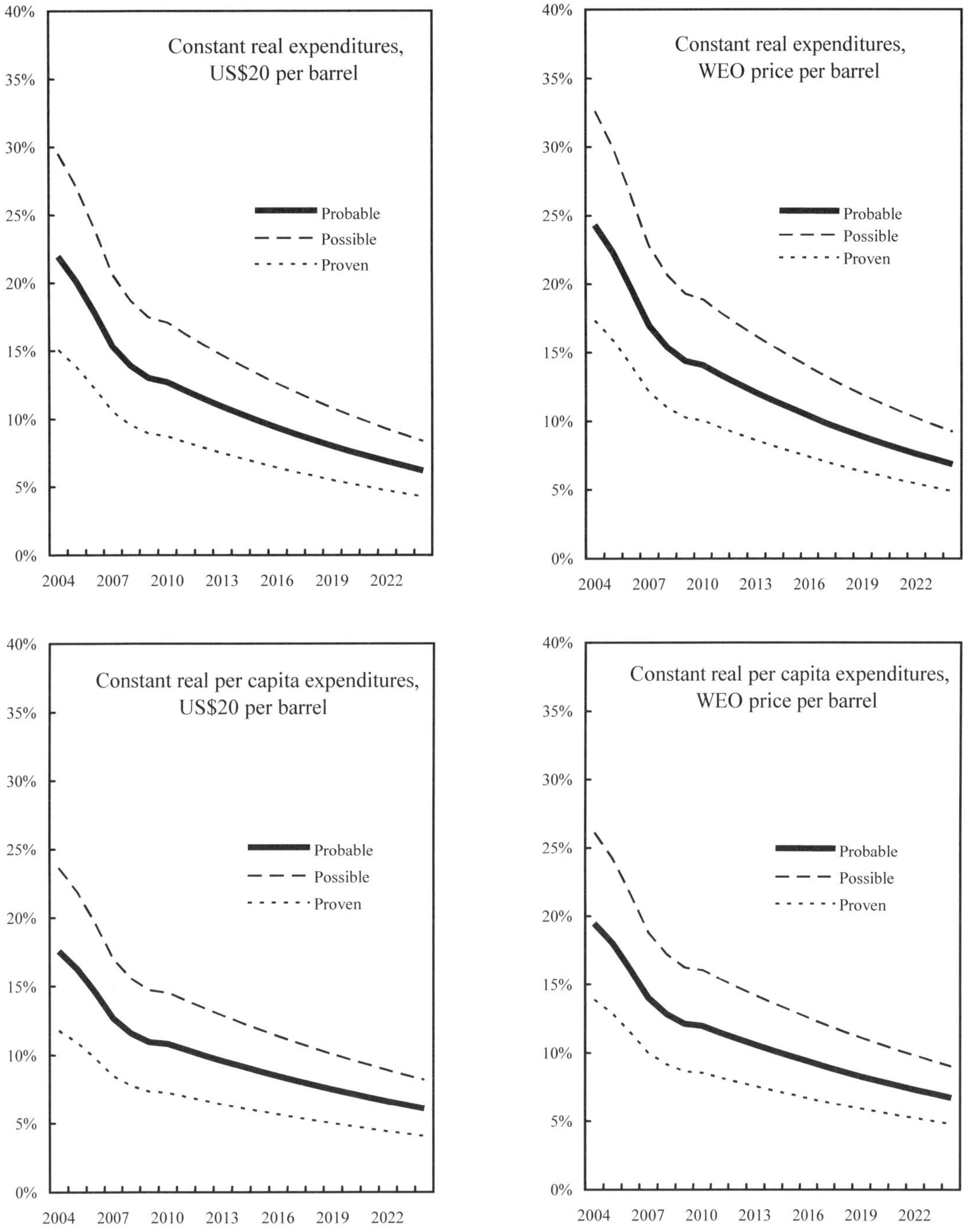

Sources: Azerbaijani authorities and IMF staff calculations.

Figure 5. Azerbaijan: Sustainable Non-Oil Deficit—Constant Wealth, 2004–24
(In percent of GDP)

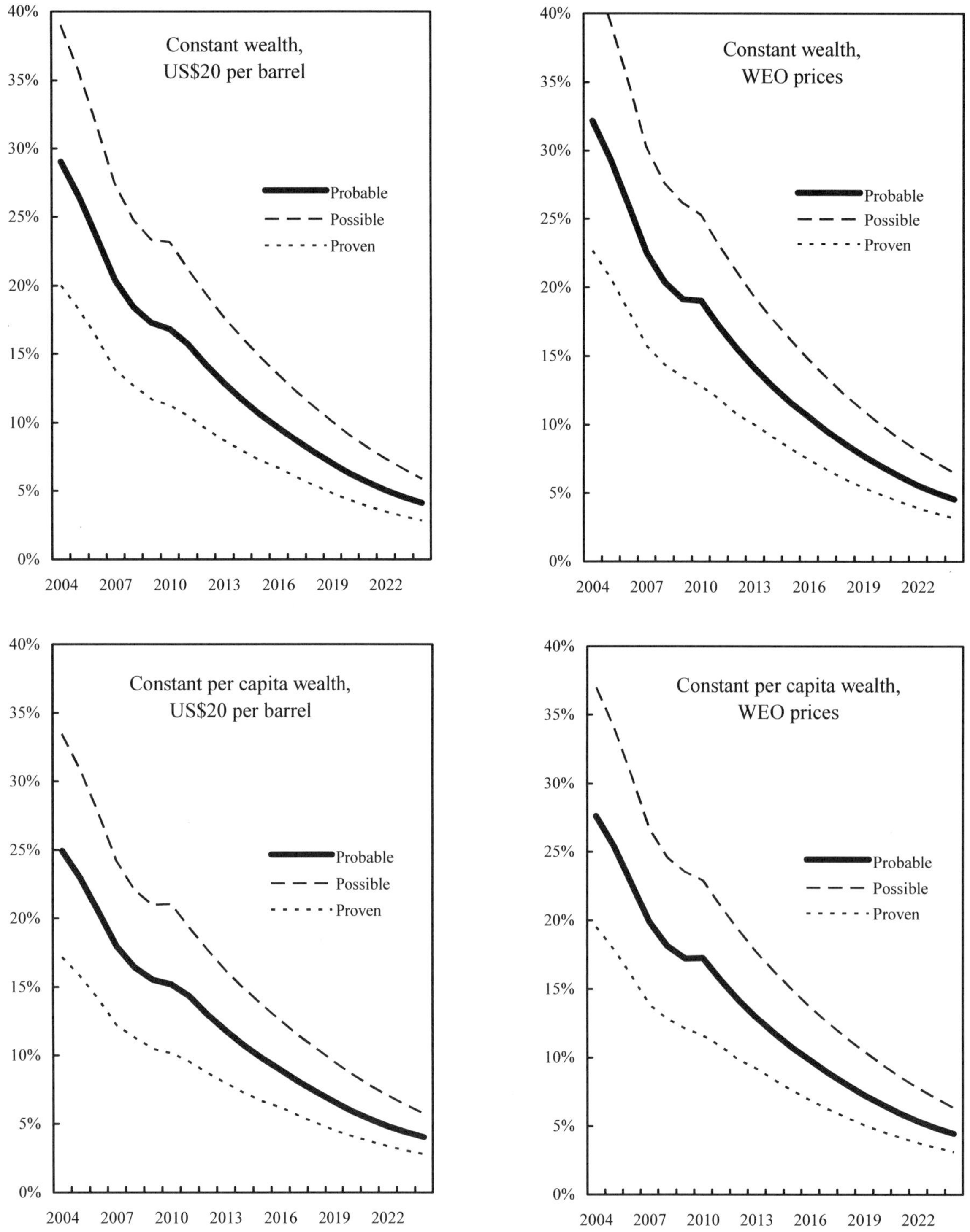

Sources: Azerbaijani authorities and IMF staff calculations.

**Table 1. Azerbaijan: Sensitivity of the Sustainable Non-Oil Deficit
to Long-Term Real Interest Rates**
(In percent of GDP) [1]

	Baseline Interest Rate (r = 5 %)			Low-Interest Environment (r = 2 %)			High-Interest Environment (r = 8 %)		
	2004	2010	2020	2004	2010	2020	2004	2010	2020
Constant real expenditure	26	15	10	20	12	7	30	18	11
Constant real per capita expenditure	22	14	9	16	10	7	27	17	11
Constant wealth	32	19	7	25	14	4	35	22	8
Constant per capita wealth	28	17	7	19	11	3	31	21	9

Source: IMF staff estimates.
Note: r is for real interest rate.
[1] Under World Economic Outlook oil prices.

Changes in the real interest rate affect the value of oil wealth and earnings on financial wealth with repercussions on the sustainable deficit ceiling (Table 1). A reduction of the long-term interest rate to 2 percentage points, other things being equal, reduces affordable spending by 6–7 percent of GDP in the near term, with a gradually smaller effect over the longer term (2–3 percent of GDP by 2010 and 1–2 percent of GDP by 2020). Not surprisingly, the opposite case is true of high real interest rates, which ease the resource constraint. The sensitivity of the sustainable ceiling to these changes is sufficiently large to warrant caution and a conservative approach to real interest rate assumptions.

Adherence to the expenditure ceiling would ensure that long-term savings objectives are met. The sustainable deficit ceiling would set a simple and transparent rule for the level of expenditures financed from oil revenue, consistent with long-term savings objectives. It would be easy to measure, and compliance with the rule could be well monitored.

However, the sustainable deficit ceiling is not meant to be an unalterable target. Rather, it is determined on the basis of a savings prescription that needs to be regularly reviewed in light of changing information and economic conditions. Like any good fiscal rule, the sustainable deficit ceiling needs to be both simple and flexible. As new and more accurate information on production patterns, natural resource reserves, or price developments become available, the government should review the appropriateness of its long-term assessments. Revisions should be made at regularly spaced intervals along with the scheduled updates for the Medium-Term Expenditure Framework (MTEF) and the Public

Investment Program (PIP), or if information becomes available indicating that the given sustainable deficit ceiling may no longer be prudent (e.g., overly optimistic production assumptions). Formal reviews should be made on the basis of prespecified rules clearly identifying changes made to relevant variables and the underlying reasons for any changes.

Finally, special considerations may lead the government to at least temporarily pursue a savings path different from the sustainable ceiling. One example is demographic financing needs associated with pension obligations related to population aging. Declining birth rates and increased life expectancies can pose significant financing requirements for future generations. In this case, precautionary savings in anticipation of future use of assets may be a prudent policy and require a tighter savings path than one permitted by the sustainable deficit ceiling. Once the financing needs materialize, the government can then increase spending, which could result in spending levels above the prescribed sustainable ceiling. For example, a wide-ranging debate in Norway on pension funding led to the postponement of consumption of oil assets in order to form reserves for future pension payments.

In Azerbaijan, a similar situation is conceivable. A sizable post–World War II birth cohort will begin to retire starting in 2010. Projected population growth rates will only moderately increase the labor force and the expected increase in the old-age dependency ratio could therefore put severe strain on the existing pay-as-you-go pension system. The government should carefully analyze whether any special considerations, such as this, would warrant a more ambitious savings trajectory than the one prescribed by the sustainable non-oil deficit ceiling.

Estimates for the long-term sustainable non-oil deficit appear not to constrain expenditure plans for the use of oil and gas revenues over the medium term in Azerbaijan. The expected oil revenue boom, while short-lived, is substantial compared to current levels of economic activity. Under existing assumptions for production plans and oil prices, the government could afford a non-oil deficit far in excess of current plans and still be able to afford the same constant real expenditures each year. However, these long-run considerations may be an insufficient guide for near-term expenditure plans.

B. Fiscal Rules for the Medium Term

The sustainable non-oil deficit paths presented in Figures 4 and 5 provide an expenditure envelope for the medium to long term; the authorities should increase actual deficits only gradually toward this ceiling. There are four main reasons for the authorities to follow a cautious strategy in increasing the non-oil deficit toward the long-run sustainable level. First, a rapid increase in deficits could exert excessive upward pressure on the real exchange rate with all its negative consequences for the non-oil sector (Dutch disease). Second, such an increase could strain the government's institutional capacity for planning, executing, and monitoring expenditures, resulting in substantial waste. Third, given the expected path of oil revenue inflow, the non-oil deficit paths of Figures 4 and 5 would require the government to borrow up to US$4 billion by 2007 (Azerbaijan's GDP is currently around US$6 billion)—a clearly risky and inadvisable strategy given the uncertainty of oil prices. Finally, the path would call for rapid increases, followed by rapid decreases, in non-oil deficits. Such changes in the fiscal impulse would not be good fiscal policy.

Given the risks stemming from an excessively rapid increase in the non-oil deficit consistent with the long-run ceiling, we add two constraints to the simulations described in the previous section to generate a medium-term strategy: a nonborrowing constraint, and a limit on the year-on-year fiscal adjustment.

A nonborrowing constraint—prohibiting borrowing against future oil revenue to finance current spending—is desirable for several reasons. These include the uncertain nature of the oil wealth, the imperfection of capital markets (e.g., limited hedging possibilities), and possible "sudden stops" of capital flows that could precipitate financial and balance of payments crises. Many examples can be found of countries that borrowed against future oil wealth, only to find that the wealth was not sufficient to service the debts.

A limit on the year-on-year fiscal adjustment is also desirable. Indeed, even under the assumption of known future prices and quantities, both Figures 4 and 5 reveal that the non-oil deficit is very volatile and declining rapidly over time.[10] Such fiscal adjustments are clearly undesirable, not only for public finances but also for the non-oil economy. Also, given the size of the adjustment, it would be politically difficult to implement. Recognizing this, market participants would start questioning the sustainability of Azerbaijan's public finances, which could trigger speculative attacks and a financial crisis. A fiscal rule designed to put a cap on fiscal adjustments would smooth fluctuations in the expenditure program, minimize Dutch disease repercussions, and also reduce waste of government resources given the limited institutional capacity of the government.

One simple fiscal rule is to set an annual fluctuation band for the non-oil deficit. Adding this constraint onto the easily understandable real constant expenditure analysis (with a nonborrowing constraint) makes this rule easy to explain, calculate, and monitor—key elements of a successful fiscal rule in practice.

---

[10]For instance, from the probable oil production scenario, a constant total oil wealth rule indicates that, for 2004, the non-oil deficit can reach 29 percent of GDP. After that, each year until 2008 it declines by around 3 percent of GDP, to reach a maximum of 18 percent of GDP.

Figure 6 shows the maximum medium-term non-oil deficit path for the baseline scenario under two approaches—constant real non-oil deficits and constant real non-oil deficits per capita.[11] This scenario is based on the probable oil production (the scenario used by oil companies when making their investment decisions), and assumes a constant price of US$20 per barrel, which is slightly more conservative than the current WEO projections. Starting from the unconstrained long-run sustainable path (Figure 4), it incorporates both the nonborrowing constraint and an annual fluctuation band for the non-oil deficit. The latter constraint is applied on the non-oil deficit relative to non-oil GDP as the best measure of the fiscal impulse. The estimated 2003 non-oil deficit is added as the starting point (15.8 percent of non-oil GDP, or 11.3 percent of total GDP). Two fluctuation bands are plotted (a maximum increase or decrease of 1 and 2 percent of non-oil GDP, leading to band widths of 2 and 4 percent of non-oil GDP, respectively).

The two fiscal rules support basically identical non-oil deficit paths over the medium term. After 2007, the constant real expenditure rule, as expected given our assumption of population growth, generates a more front-loaded path of fiscal deficits. Under the 2 percent band width limit, for example, the simulations on the basis of the constant real expenditure rule initially yield a deficit of about 0.5 percentage point of non-oil GDP greater than the constant real expenditure per capita rule in 2008, followed by 1.4 percentage points in 2009. The difference in the deficit paths between the two scenarios declines gradually over time, and the two paths intersect in 2035, after which the deficit path under the former rule is lower than the path under the latter rule.

The baseline scenario summarized above supports an increase in the non-oil deficit over the medium term in the range of 1–2 percent of non-oil GDP (i.e., the maximum allowable increase in each scenario). The 2 percent band, by definition, generates a smoother path of fiscal deficits over the medium term relative to the 4 percent band. However, equally important, the nonborrowing constraint becomes binding for the 4 percent band in 2006, whereas it is never binding for the 2 percent band. (The fact that the nonborrowing constraint does not bind is a desirable feature given the uncertainty surrounding all these estimates, i.e., a financial "safety margin" remains.)

---

[11]These approaches are judged to be preferable to the constant wealth strategies, as they would be politically easier to explain and defend.

Figure 6. Azerbaijan: Medium-Term Sustainable Non-Oil Deficit Ceiling (Includes a Nonborrowing Constraint (NBC) and a Limit on the Non-Oil Deficit Volatility), 2003–24
(In percent of GDP)

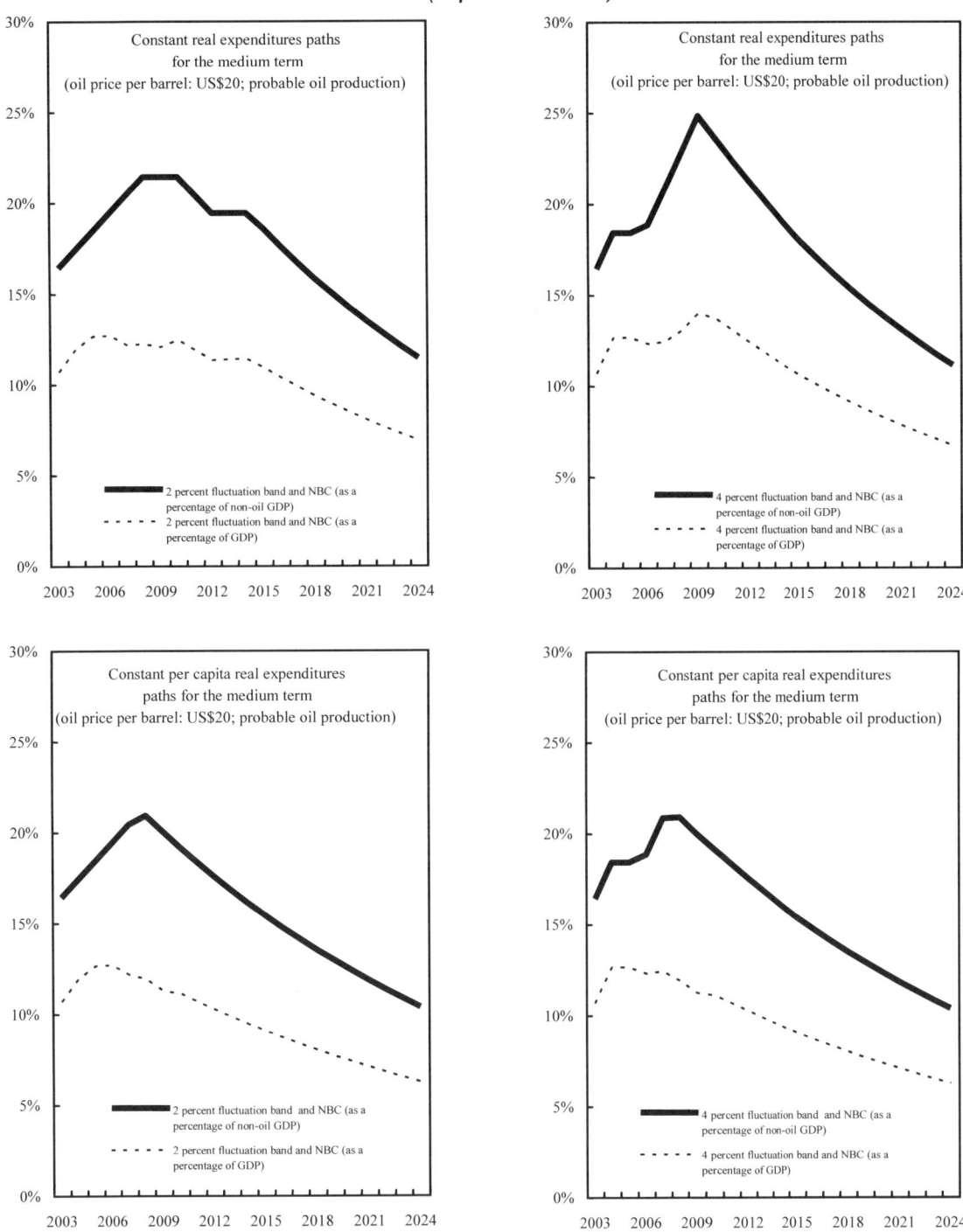

Sources: Azerbaijani authorities and IMF staff calculations.

23

In addition, sensitivity analysis suggests that, to be on the cautious side, the increase in the annual non-oil deficits over the medium term toward the long-run sustainable path should be closer to the lower bound of the 1–2 percent range suggested above. Figure 7 provides non-oil deficit paths for the constant real expenditure rule based on various oil prices (i.e., US$19, US$21, and US$18 per barrel).[12] Based on a price of oil of US$18 per barrel, while the financial assets already accumulated in the oil fund would allow the government to finance, for 2004, a non-oil deficit of 2 percent of non-oil GDP higher than the expected deficit for 2003 (bringing the deficit to 17.8 percent of non-oil GDP in 2004), the nonborrowing constraint would become binding in 2005, and the deficit would have to decline to about 17 percent of non-oil GDP in 2006. Thus, following the 2 percent limit under the US$18 per barrel scenario would be undesirable, as it would deplete the financial assets of the government by 2006. On the other hand, if the 1 percent limit was instead imposed on the non-oil deficit, the nonborrowing constraint would not bind, and the non-oil deficit would be constant or gradually increasing through 2008 (reaching 20 percent of non-oil GDP), before starting to decline in line with the long-run sustainable non-oil deficit path.

C. Policy Options in the Medium Term

Once a desirable short- to medium-term fiscal ceiling is determined, another challenge for the government remains: what is the optimal use of the non-oil deficit? What is the proper mix between tax reductions and expenditure increases? And what types of taxes should be cut or expenditures increased?

In deciding on the optimal use of the non-oil deficit, country-specific features need to be taken into account. In addition, traditional policy instruments will have to adapt to the new challenges of large asset flows in order to cope with the implications for economic stability. Policies for the medium term also have to be well coordinated and comprehensive to ensure consistency with non-oil sector development. Finally, chosen policies will have to be sufficiently flexible (e.g., prioritized) to allow an adequate response to changing economic conditions.

The main burden of maintaining economic stability will be on fiscal policy, since monetary instruments have only limited capacity to manage resource inflows. Given a modest holding of government securities and a thin market for central

---

[12]See Table 2 in Appendix 1 for the first five years of data for the non-oil deficit path under various oil prices and fluctuation bands.

Figure 7. Azerbaijan: Sensitivity Analysis of the Medium-Term Sustainable Non-Oil Deficit Ceiling, 2003–24

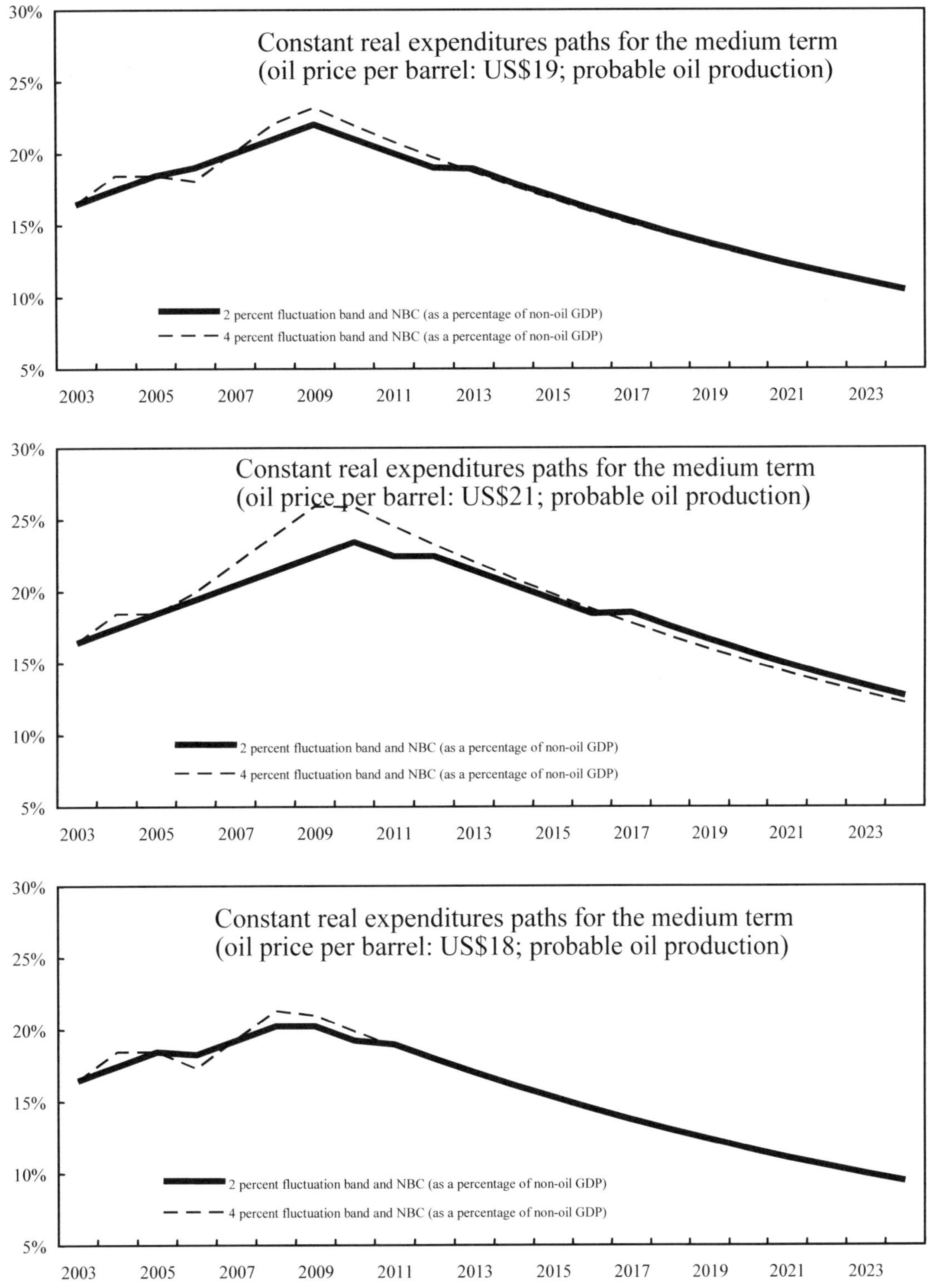

Sources: Azerbaijani authorities and IMF staff calculations.

Box 2. Monetary Policy Response to Natural Resource Booms

Experience with natural resource booms suggests that some degree of real exchange rate (RER) appreciation is inevitable, and actually desirable to effect the reallocation of factors of production in the economy necessary to accommodate these booms. The response of monetary policy has important implications for the channels through which this RER appreciation takes place.

A resource boom typically raises domestic absorption and, therefore, the demand for money in real terms. If monetary policy does not accommodate this increase in money demand through an expansion of money supply, the result will be an excessive appreciation of the real exchange rate. On the other hand, if the central bank resists a nominal exchange rate appreciation, the inevitable RER appreciation will take place through higher inflation, thus undermining the main goal of monetary policy—i.e., the maintenance of domestic price stability.

So while the central bank should accommodate, at least in part, the increase in real money demand, the key will be to strike a balance between price stability and nominal appreciation. Prudent fiscal policies are essential for this balance to be reached, particularly in the case of countries with underdeveloped domestic financial markets, such as Azerbaijan. Experience from Asia, for example, suggests that open market operations have proved inadequate to stabilize monetary growth, and thus inflation, during periods of particularly severe disturbances (Tseng and Corker, 1991). This reflected a variety of factors, but most importantly included an inadequate development of markets and instruments for open market operations.

In the case of Azerbaijan, sterilization of excessive spending of the oil boom may be difficult to sustain, in view of the Azerbaijan National Bank's (ANB) limited holding of securities and a thin market for ANB bills. Under these conditions, the effectiveness of open market operations will be constrained for the foreseeable future. It is therefore imperative that fiscal policy remain prudent and consistent with the maintenance of macroeconomic stability, and that it not lead to an excessive appreciation of the real exchange rate. There is also a need to strengthen coordination of macroeconomic policies between the Ministry of Finance and the ANB.

bank bills, sterilization of excessive spending of the oil boom may be difficult to accomplish (Box 2). By directly controlling the injection of oil revenue into the economy, fiscal policy is therefore the key tool for macroeconomic management.

Strengthening fiscal policy will require a less fragmentary approach to managing the use of oil revenue and a careful analysis of the overall implications for domestic demand. Currently, oil revenue is managed by different government agencies (the state budget and SOFAZ) and the use of resources is not systematically coordinated. While assets from SOFAZ are primarily directed toward capital projects, revenues accruing from SOCAR's domestic operations are perceived as a general government finance source. In addition, while the government seeks to use the oil fund as an instrument for saving oil wealth, completely separate arrangements—independent of the oil fund—are being made for stabilizing the flows of oil revenue to the state budget. This treatment clouds the true dependency of government operations on oil revenue and makes coherent demand management difficult. By treating all oil revenues as a single source of financing, the government could better manage the overall impact of its use on the economy. In conjunction, the government should develop and maintain a model for projections of oil and gas revenues for planning purposes.

Strategic planning and enhanced coordination of macroeconomic policies will be crucial to accomplish this goal. In particular, the annual budget should be firmly embedded within a sound medium-term expenditure strategy that balances the needs of macroeconomic stability and non-oil sector growth with expenditure priorities from the Poverty Reduction Strategy Paper (PRSP) and the government's investment program. The appropriateness of fiscal and monetary policy will have to be regularly reassessed (e.g., quarterly), discussed in a broad government forum, and realigned if necessary. In addition, government institutions, such as the public investment unit, will have to be strengthened as discussed below.

Equally important for managing the macroeconomy and non-oil sector growth is carefully planning the content and composition of overall expenditures. As government spending affects domestic demand and influences private sector activities, the specific use of oil revenue greatly influences economic stability and non-oil sector growth prospects.[13] Therefore, when designing its medium-term strategy, the government should thoroughly analyze the composition of its overall spending plans:

- *Current versus capital spending?* Since oil revenue is an exhaustible source of financing, it is generally preferable to direct spending toward capital projects. Financing needs for investments are by nature limited and not permanent as is the case for current expenditures. In addition, current spending (e.g., public wages) often directly fuels domestic consumption, as consumers demand domestically produced goods and services. On the other hand, investment projects with high import content may have only a moderate direct effect on domestic demand.

- *What types of capital investments should be undertaken?* A key role of the government is to generate conditions in support of private sector development, including a well-designed and reliable physical infrastructure. Thus, capital investments should target basic infrastructure needs, such as the

---

[13]Azerbaijan has little nonconcessional government debt, and therefore use of oil revenues for early repayment of nonconcessional debts is not a viable option.

reliable provision of energy and water, an efficient transport and communications network, particularly in regions outside the capital city, and improvements in health and education services. Such capital expenditures will have a direct positive impact on the competitiveness of the non-oil sector, stimulate regional development, and help offset the negative effects of an appreciated real exchange rate.

• *What are the implicit commitments contained in capital expenditures?* Large investment projects not only require expenditures over several years, but also bring with them substantial future maintenance costs. These costs are often underestimated and have led to a significant waste of resources when maintenance costs could no longer be afforded. In this context, the adoption of a notional maintenance fund could be advisable (Box 3). Similarly, expansions in pension and social programs can have significant long-term implications that could erode the government's ability to manage the use of its oil revenue assets.

One way increased non-oil deficits can strengthen the private sector is through reductions in tax rates. Instead of using oil revenue exclusively for additional expenditures, the government could reduce the tax burden within the country by lowering taxes on the non-oil sector. This policy has the advantage of potentially reaching a large share of the population and thus broadly distributing the benefits from oil wealth. In addition, a lower tax burden can ease competitiveness pressures from a higher real exchange rate and offset some of the potentially damaging effects of increased oil-revenue spending.

However, the government should carefully weigh the benefits and costs before undertaking any far-reaching tax policy changes. First, tax policy should be evenly applied by removing widespread exemptions. Significant uncertainty about the government's oil wealth warrants caution. Reductions in tax rates are likely to be permanent, since tax cuts are usually hard to reverse. An erosion of the domestic tax base could therefore backfire if the value of oil assets has been overestimated or assets are depleted faster than anticipated.

Finally, addressing medium-term challenges for fiscal policy design requires significant capacity building. In particular, the government needs to increase its ability in fiscal policy analysis and project appraisal in order to effectively implement a viable medium-term fiscal policy. Improved fiscal policy analysis will strengthen the government's ability to assess current developments, identify the appropriate fiscal stance, and react when necessary to changes in the short-term macroeconomic environment. In parallel, the government needs to devote additional human resources to strengthening macroeconomic policy formulation and building a viable public investment unit for expenditure planning and project

---

Box 3. Establishing a Notional Investment Maintenance Fund

A responsible investment strategy should not only incorporate immediate project costs but also proactively save for long-term maintenance costs. These recurring costs are often vastly underestimated and have led to the significant waste of assets in many countries with resource booms. But even in cases where resources were available, maintenance expenditures have been deferred to allow for the launch of prestigious new projects. One way to avoid this problem would be to identify already committed assets for future capital maintenance purposes. This could be done by creating a notional investment maintenance account. Whenever a new investment project is undertaken, this account would be credited with the present value of future maintenance costs of the project. At the same time, the stock of available assets in the oil fund for financing new projects would be reduced by the same amount. In the course of budget preparations—after the government has set a limit on overall oil fund financing—the government would first have to meet maintenance costs for existing capital projects from the maintenance fund. Only afterwards, when all maintenance needs are met, could it commit new assets from the remaining (nonmaintenance) asset pool for new investment projects. This arrangement would ensure that (1) sufficient resources are available for maintenance, (2) maintenance costs for existing projects are met before new projects are initiated, and (3) no resources are committed in excess of the sustainable expenditure ceiling. The maintenance fund would also demonstrate to the public the amount of oil and gas assets that have already been committed, and what is truly available for new projects.

evaluation. This will allow the government to prepare prioritized expenditure plans, review the productivity of individual projects, and assess consistency with the overall policy objectives of macro-stability and non-oil sector growth. In the absence of adequate institutional capacity, the government may run the risk of undertaking projects with low social returns, leading to a waste of resources, as has been documented above.

6 Conclusion

In the near future, Azerbaijan is expected to benefit from a substantial, but short-lived, oil- and gas-related revenue windfall. Even under conservative assumptions, revenues accruing to the country are expected to average around US$800 million during the period 2003–07 and over US$2 billion per year during the period 2008–24, compared to 2002 GDP of just over US$6 billion. Because few countries have been successful in managing natural resource wealth of this relative magnitude, the government faces a key and immediate challenge: managing this short-lived natural resource wealth in such a manner as to avoid the pitfalls of Dutch disease and ensure the simultaneous development of the non-oil sector.

This paper aims to provide a broad policy agenda for the government for managing this natural resource wealth. The key policies and recommendations in the paper are as follows.

A. Institutional Arrangements and Capacity

- Consolidate oil revenue management and treat all oil revenue as one source of financing.

- Develop and maintain a model for long-term projections of oil and gas revenues.

- Develop institutional capacities for project selection, monitoring, and evaluation, including the establishment and development of a project appraisal department as well as capacity building in fiscal policy analysis.

B. Level of Expenditures

- Set expenditures of oil and gas revenues consistent with a long-term savings objective of conserving assets for the future, particularly given the short-lived nature of the windfall. The goal should be to ensure constant real expenditures out of oil wealth.

- Use the concept of a sustainable non-oil deficit to provide an expenditure ceiling for the use of oil assets that is consistent with this long-term savings objective. Under the baseline scenario for oil and gas reserves and conservative assumptions for the price of oil, substantial non-oil deficits are affordable until 2010, with subsequent steadily declining non-oil deficits.

- Avoid large fluctuations in the non-oil deficit.

- Revise the estimate of the sustainable non-oil deficit in light of new information. The appropriateness of the sustainable non-oil deficit should be reviewed at regular and sufficiently spaced intervals, based on updated information on oil and gas reserves, production patterns, and price developments.

- As the sustainable non-oil deficit provides only an expenditure envelope for the medium term, do not increase expenditures to this ceiling in the near future. This would not be advisable given the macroeconomic implications of excessive growth in spending. In particular, a rapid increase in expenditures consistent with this ceiling could exert substantial upward pressure on the exchange rate with all its negative consequences for the non-oil sector. It could also strain the government's institutional capacity for planning, executing, and monitoring expenditures, resulting in substantial waste.

- Do not borrow against future oil revenue to finance current spending.

- Take macroeconomic stability considerations into account when deciding how much oil revenue to spend in the medium term. Strengthened coordination between the Ministry of Finance and the Azerbaijan National Bank will be imperative.

C. Composition of Expenditures

- Revenues should be utilized primarily for investment rather than consumption. Expenditures on physical and human capital will provide a solid foundation for the future growth of the country, while excessive current consumption could have a potentially destabilizing impact in the short term. Capital expenditures have the added advantage of a substantial import

31

content, providing an automatic means of sterilizing part of the substantial foreign exchange inflows associated with the oil windfall.

- Capital investment should target the building and maintenance of a well-designed physical infrastructure necessary for improving the competitiveness of the non-oil sector, including the reliable provision of energy and water, an efficient transport and communications network, and improved education and health services, particularly in the regions outside the capital city.

- A notional investment maintenance fund should be established for meeting recurrent costs associated with physical infrastructure projects. This would increase transparency of already committed resources and ensure proactive savings for long-term maintenance costs.

- Reductions in tax rates could be an alternative to increased expenditures, with the direct positive impact on competitiveness offsetting, at least in part, the negative effects of real appreciation.

Political pressures for excessive and speedy expenditures of oil wealth are inevitable. Successful long-run development of the economy will require that the government withstand such pressures. But this will not be easy. The government will need to demonstrate to the population not only that oil wealth is being saved for future generations but that it is also being used to effectively benefit the current population of Azerbaijan. The policies recommended above—focusing on infrastructure development and protecting non-oil competitiveness—should help generate new employment opportunities and meaningful economic growth. If the government can succeed in doing this, and also succeed in explaining to the population the dangers—not just to future generations but to the current population of Azerbaijan as well—of excessively rapid expenditures of oil wealth, Azerbaijan may succeed where so many other oil-producing countries have failed: it may manage to use its oil wealth to help develop the non-oil sectors of its economy.

1 Oil and Economic Development in Indonesia[14]

Indonesia's oil industry is one of the world's oldest. Indonesia ranks fifteenth among world oil producers, with about 2.4 percent of world oil production. The country has a mixed economy in which the government, in addition to the regulation and supervision of the economy, is engaged directly in economic activities through state-owned enterprises operating in various sectors.

Between 1960 and 1966, the country suffered from hyperinflation, and GDP grew at an average rate of only 1.8 percent per year. In 1966, the government started the implementation of an economic policy program ("New Order") designed by a team of presidential economic advisors. Stabilization was achieved soon thereafter in 1971 with 4 percent inflation and 6 percent GDP growth. In 1973, oil exports accounted for only around a third of total exports because of the country's richness in natural resources (rubber, coffee, timber). International reserves grew rapidly after the first oil boom in 1972–78, and the windfall oil revenues of 1973–78 allowed the authorities to increase spending on development. Around half of mining value added was used to finance public investment, one third was used to reduce the trade and nonfactor services deficit, and the rest was spent on consumption.

The rapid growth of international reserves, together with high domestic spending, contributed to a sharp real exchange rate appreciation, and many non-oil sectors, such as rubber and manufacturing, started to experience difficulties. The government regarded increasing dependence on oil revenues as risky in light of uncertain prospects for oil prices and realized that future growth had to come from labor-intensive exported goods. In 1978, the government decided that devaluation of the domestic currency would help restructure the economy, to make it less reliant on oil and to move toward manufactures and non-oil exports. The devaluation of the currency by 50 percent was followed by inflation of 22 percent in 1979. The devaluation was generally regarded as successful, as

---

[14]Sources: Dornbusch and Helmers (1987); and Alan Gelb and Associates (1988).

Table 2. Azerbaijan: Constant Real Non-Oil Deficit Paths Under Various Oil Prices and Fluctuation Bands
(In percent of non-oil GDP)

	US$18 Oil Price		US$19 Oil Price		US$20 Oil Price		US$21 Oil Price	
	2-Percent Band	4-Percent Band	2-Percent Band	4-Percent Band	2-Percent Band	4-Percent Band	2-Percent Band	4-Percent Band
2004	17.4%	18.4%	17.4%	18.4%	17.4%	18.4%	17.4%	18.4%
2005	18.4%	18.4%	18.4%	18.4%	18.4%	18.4%	18.4%	18.4%
2006	18.2%	17.3%	19.0%	18.1%	19.4%	18.9%	19.4%	19.9%
2007	19.2%	19.3%	20.0%	20.1%	20.4%	20.9%	20.4%	21.9%
2008	20.2%	21.3%	21.0%	22.1%	21.4%	22.9%	21.4%	23.9%
2009	20.2%	21.0%	22.0%	23.2%	21.4%	24.9%	22.4%	25.9%

Source: Staff estimates.

manufactured exports doubled during 1978–79 and the non-oil trade balance improved. The reason for devaluation was not balance of payments troubles—reserves coverage was at four months of imports. The aim was to help the relatively labor-intensive non-oil traded sectors.

The second oil boom raised Indonesia's mining sector revenues again. The government increased spending once more, but the absorption of windfall oil revenues was much below the level of expected oil income, and foreign aid and part of revenues were saved. This differed from the approach during the first oil boom. Oil prices started to fall in 1981, and due to a rapidly growing trade imbalance, the current account turned into a large deficit. Capital inflows were insufficient to finance the high trade deficit and foreign exchange reserves started to fall. The authorities decided to devalue the currency again to stop private capital outflows in the short term and to improve the non-oil trade balance. In 1983, the domestic currency was devalued by around 50 percent, and for the second time in five years, relative prices of traded goods and nontraded goods changed sharply. Fiscal policy was tightened and was supportive of the devaluation. In mid-1983 more than $10 billion in capital-intensive public projects, amounting to almost 12 percent of GDP, were canceled or postponed. This sharp reduction in government spending allowed the government to implement an expenditure-switching policy from industry to infrastructure and social sectors. In 1984, the authorities introduced a simplified tax code with a rudimentary form of a value-added tax, which is easier to administer and monitor for the non-oil sector. Immediately following the devaluation, the authorities liberalized the financial system to create incentives for lending, and increase competition and greater mobilization of domestic savings. This devaluation proved successful too—during 1983–85 non-oil and manufactured exports increased, non-oil imports fell, foreign exchange reserves strengthened, and the overall government budget returned to balance.

Indonesia's experience with oil windfall management stands out as relatively successful compared to other oil-exporting countries. Three key factors contributed to this success: oil was not the only source of export earnings and exports of other commodities were generating considerable income; the authorities did not rely on oil sector revenues alone and tried to diversify the economy—the country was a strong non-oil exporter during the periods of the oil booms; and the Indonesian government adapted macroeconomic policies to the changing external environment.

2 Oil and Economic Development in Nigeria[15]

Nigeria has an abundance of hydrocarbon resources. It is the thirteenth largest oil producer in the world, the third largest oil producer in Africa, and the most prolific oil producer in sub-Saharan Africa. Prior to 1960, agriculture was the dominant sector in the Nigerian economy and the country was a major producer of cocoa and palm products. Oil production in Nigeria started in 1958 and increased over time to reach the export of 2 million barrels of oil per day by 1972.

With the first oil boom of 1972–78, Nigeria's terms of trade increased three times; international reserves increased almost tenfold between 1973 and 1974. Oil revenues accounted for almost 85 percent of the country's total exports and around 60 percent of federal government revenues in 1973. At this stage, the government faced the question of how to use such vast unplanned revenues. The fiscal authorities ignored the risk of future reversal of the current favorable conditions and chose to spend these revenues by undertaking massive domestic investment projects. Public capital spending accelerated rapidly, absorbing more than the total increase in 1970–76 oil revenues, resulting in a large budget deficit, which was financed with the use of reserves accumulated in 1973–74 and monetary expansion. These policies resulted in inflation—prices increased by 22 percent and, with a mainly fixed exchange rate, the real exchange rate appreciated strongly.

The government was not successful in diversifying the economy, particularly as specific policies further negatively affected the once strong agriculture sector. Production of major agricultural export crops shrunk by half from 1964 to 1978, partly because the government created commodity boards to stabilize crop prices and taxed farmers by paying them substantially less than world prices. Nigeria became a net importer of agricultural products in 1975. The government

---

[15]Sources: Dornbusch (1993); Alan Gelb and Associates (1988); Mered (1997); and IMF (2002).

responded to the difficult economic situation by expenditure cuts in 1978, but did not address the issue of the overvalued real exchange rate.

The second oil boom saved the government from undertaking further painful adjustments. Nigeria's terms of trade increased by 25 percent and 40 percent in 1979 and 1980, respectively, and the international reserves position strengthened significantly. However, the Nigerian government did not take into account the lessons of the past. In light of the increasing oil revenues, fiscal constraints were relaxed and expenditures rose by 65 percent in 1980, to resume the suspended construction projects and to undertake new ones. However, the second oil boom did not last long; oil export receipts halved between 1980 and 1982, and this expansionary fiscal policy resulted once again in large fiscal deficits by 1982. Foreign exchange reserves fell sharply and the real effective exchange rate appreciated by 125 percent compared to its 1976 level. Inflation reached 60 percent during 1980–83. The government introduced restrictive quantitative controls and import quotas on goods and services that hurt the manufacturing sector. In addition, payments arrears on foreign debt were accumulated, adversely affecting Nigeria's credibility in international capital markets. At this point, the government approached creditors to prolong existing loans and to get new financing. By the end of 1983, the Nigerian economy was in trouble again.

Nigeria failed to use its oil wealth for the benefit of its people during the boom years. Experience in Nigeria shows that the high level of expenditures during oil boom periods was difficult to reverse after price falls, thus resulting in widened fiscal deficits. Fiscal volatility adversely affected the economy through appreciating real exchange rates. The authorities spent the oil income mainly for domestic investment and consumption. Any savings of oil revenues was short-lived; revenues were saved only immediately following the surge in windfall income and were then subsequently spent quickly. The large public investment projects did not succeed because of constraints in the implementation process. Investments in the industry sector failed to generate the much-needed non-oil exports and the authorities did not manage to diversify the economy during the windfall decade. The decision to adjust to shrinking oil revenues through trade restrictions rather than through devaluation had a ruinous impact on the economy. In addition, heavy and long dependence on oil revenues resulted in a narrowing of the non-oil tax base and inefficient tax administration, which played a negative role in the country's macroeconomic performance throughout the 1980s and 1990s, as oil prices fluctuated.

3 Oil and Economic Development in Mexico[16]

Mexico is the world's fifth largest oil producer and its tenth largest oil exporter. Mexico began to export oil in 1911, and its oil output expanded at an average annual rate of 6 percent between 1938 and 1971. Extensive oil discoveries in the 1970s increased Mexico's domestic output and export revenues.

Although the Mexican economy maintained a rapid growth rate during most of the 1970s, it was progressively undermined by a combination of fiscal mismanagement and an overvalued real exchange rate, resulting in the sharp deterioration of the investment climate. In the mid-1970s, the government planned large public sector investment programs in industry, agriculture, and transportation. This expansionary fiscal policy, together with expansionary monetary policy, the postponement of crucial tax reforms, and a fixed exchange rate contributed to large balance of payments disequilibrium and intensified capital outflows. In 1976, the government devalued the peso by 45 percent. In the same year, Mexico agreed with the IMF on a stabilization program aimed at lowering inflation, building up reserves, and achieving macroeconomic stability. Oil discoveries in the south of Mexico in 1978 and a sharp increase in the world price of oil in 1979 greatly affected the country's economic outlook. Private capital started to flow into the country, financing from the IMF was no longer needed, and the reform program was abandoned.

The improved terms of trade in 1979–80 brought windfall oil revenues and allowed the government to continue implementing an expansionary fiscal policy. Moreover, the government borrowed abroad against future oil earnings to further boost expenditures. Public investment increased and reached 30 percent of GDP in 1981. This growth was associated with a substantial increase in imported capital and intermediate goods. However, oil revenues were not sufficient to finance the large increase in imports and external imbalances were financed by foreign borrowing. The budget deficit rose, the current account

---

[16]Source: Dornbusch and Helmers (1987).

deficit widened, and the real exchange rate was allowed to appreciate. Oil became the economy's most dynamic growth sector, and the country's dependence on income from the export of oil increased. The share of oil in total exports rose from 15 percent in 1976 to 78 percent in 1983. Government tax revenues were now heavily dependent on international oil price movements. When oil prices fell in 1981, the government decided not to cut prices for Mexican oil for several months and the volume of oil exports fell sharply. In 1982, the budget deficit reached 15 percent of GDP. In the same year, commercial banks refused to roll over government loans. In August 1982, Mexico suspended its international debt payments after falling oil prices made it impossible for the government to repay foreign loans. Around $30 billion of capital fled the country. The debt crisis led to currency devaluations and hyperinflation.

Mexico's experience with oil revenue management is a good example of how the existence of abundant natural resources can create a false sense of security. Even windfall resources from oil during the skyrocketing oil price period could not sustain overly expansive public spending, and the country faced the painful need for adjustment later on. In fact, the discovery and exploitation of oil resources gave a false sense of security to the authorities and made them postpone the needed correction of the real exchange rate, balancing of the budget, and implementation of various structural reforms.

4

Calculation of the Sustainable Non-Oil Deficit

The paper discusses two long-term strategies for the use of oil assets. The first is based on a fixed expenditure rule, which—together with a permanent financing condition constraint—pins down the time path for the sustainable non-oil deficit. It assumes that the government will ensure that a predetermined expenditure path can be permanently financed out of an accumulated financial asset stock. The second strategy aims to preserve oil wealth indefinitely and thus creates a permanent income and expenditure stream. This strategy requires adherence to a long-term savings path. The level of available resources for consumption, that is, the sustainable non-oil deficit path, derives as a residual.

The presented scenarios examine the long-term implications of these two strategies. The calculations explore the direct effect of changes in crucial parameters such as the oil price of the real interest rate, but do not model their interactions with other variables such as the growth or inflation rate. The following two sections discuss the computation of the sustainable deficit under the two strategies.

A. Constant Real Expenditure or Constant Real Per Capita Expenditures

The government guarantees financing of constant expenditures in real terms or in real per capita terms (alternative scenario):

$$E_t = E_{t-1}(1+p) = E_0(1+p)^t,$$

where p denotes either the long-term inflation rate or the population cum inflation growth rate. E_0 denotes the initial level of expenditures. The sustainable non-oil deficit is defined as the corresponding financing need for this consumption path. In order to secure this financing stream a sufficiently large stock of financial assets FW needs to be saved by period T when the revenue stream ceases ($R_T = 0$), so that expenditure financing can continue. The permanently fundable level of real expenditure from FW_T is the real interest earnings on the assets stock:

$$E_t = FW_t \, i \text{ for all } t \geq T.$$

The formation of the financial wealth FW_t follows the dynamic process

$$FW_t = FW_{t-1}(1 + i + p) + (R_t - E_0 p^t) \, \alpha \, (1 + i + p),$$

where the fraction α denotes interest earned on within-year flows and was set to $\alpha = \frac{1}{2}$, assuming a constant rate of within-period net flows. At a given revenue stream and interest rate, the level of financial assets is determined by the initial level of expenditures. Thus the maximum E_0 that can be chosen has to be sufficiently low to generate a financial wealth stock by period T, which allows permanent financing of the expenditure path, or

$$E_0(1 + p)^T = F_T \, r = E_T.$$

In order to solve for the maximum sustainable deficit path, the following iterative process can therefore be applied:

1. Choose a level for E_0;

2. Determine $FW_T(E_0)$;

3. Determine $E_T(1+p)$;

4. Check whether the permanent financing condition is met.

If $E_0(1 + p)^T < F_T(E_0) \, r_T$, increase E_0 and go back to step 2.

If $E_0(1 + p)^T > F_T(E_0) \, r_T$, decrease E_0 and go back to step 2.

B. Constant Real Wealth and Constant Per Capita Wealth[17]

This strategy aims to preserve oil wealth in order to generate a permanent income flow. In particular it requires that total oil wealth TW (or per capita wealth) remains constant for all periods and thus enables a permanent income stream. Total oil wealth TW is defined as the value of physical oil assets OW_t in

---

[17]For a detailed derivation see, for instance, Davoodi (2002).

the ground plus the value of financial assets FW_t created from savings out of oil revenue:

$$TW = OW_t + FW_t .$$

The stock of oil wealth in the ground is given as the present discounted value of future revenue streams and declines over time as reserves in the ground are depleted:

$$OW_t = \sum_{k=1}^{T} \frac{R_{t+k}}{(1+i)^k} .$$

Financial wealth in period t is determined by interest earnings and additions from new savings S_t:

$$FW_t = (FW_{t-1} + S_t)(1+i) .$$

As oil wealth gradually declines over time, financial wealth and thus savings have to adjust to ensure that total wealth (per capita wealth) remains constant. Therefore the condition $TW_t = TW_{t-1}$ implies that

$$OW_{t-1} - OW_t = FW_t - FW_{t-1} .$$

From above it is known that

$$OW_t - OW_{t-1} = i\, OW_{t-1} - R_t$$

and

$$FW_t - FW_{t-1} = i\, FW_{t-1} + S_t(1+i),$$

so that it is now possible to solve for the required savings path S_t:

$$S_t = 1/(1+i)\{R_t - i(OW_{t-1} + FW_{t-1})\}.$$

The path of permissible expenditures is then given by $E_t = R_t - S_t$, which represents the sustainable expenditure ceiling.

References

Azariadis, Costas, and Allan Drazen, 1990, "Threshold Externalities in Economic Development," *Quarterly Journal of Economics,* Vol. 105 (May), pp. 501–26.

Barnett, Steven, and Rolando Ossowski, 2002, "Operational Aspects of Fiscal Policy in Oil-Producing Countries," IMF Working Paper 02/177 (Washington: International Monetary Fund).

Corden, W.M., 1992, *International Trade Theory and Policy: Selected Essays of W. Max Corden* (Aldershot, Hants, England; Brookfield, Vermont: Edward Elgar).

Davoodi, Hamid, 2002, "Assessing Fiscal Vulnerability, Fiscal Sustainability, and Fiscal Stance in a Natural Resource Rich-Country," in *Republic of Kazakhstan—Selected Issues and Statistical Appendix,* IMF Staff Country Report No. 02/64 (Washington: International Monetary Fund).

Dornbusch, Rudiger, ed., 1993, *Policymaking in the Open Economy: Concepts and Case Studies in Economic Performance* (New York: Oxford University Press for the World Bank).

———, and F. Leslie C.H. Helmers, eds., 1987, *The Open Economy: Tools for Policymakers in Developing Countries* (New York: Oxford University Press for the World Bank).

Fasano-Filho, Ugo, 2000, "Review of the Experience with Oil Stabilization and Savings Funds in Selected Countries," IMF Working Paper 00/112 (Washington: International Monetary Fund).

Gelb, Alan H. and Associates, 1988, *Oil Windfalls: Blessing or Curse?* (New York: Oxford University Press for the World Bank).

Gylfason, Thorvaldur, 2001, "Lessons from the Dutch Disease: Causes, Treatment, and Cures," Institute of Economic Studies Working Paper No. W01:06 (Reykjavik: University of Iceland).

International Monetary Fund, 2002, "Nigeria—Selected Issues and Statistical Appendix," SM/02/371 (Washington).

Karl, Terry Lynn, 1999, "The Perils of the Petro-State: Reflections on the Paradox of Plenty," *Journal of International Affairs,* Vol. 53 (Fall), pp. 31–48.

Mered, Michael, 1997, "Nigeria," in *Fiscal Federalism in Theory and Practice*, ed. by Teresa Ter-Minassian (Washington: International Monetary Fund).

Persson, Torsten, and Guido Tabellini, 1994, "Is Inequality Harmful for Growth?" *American Economic Review*, Vol. 84 (June), pp. 600–21.

Sachs, Jeffrey D., and Andrew M. Warner, 1995, "Natural Resource Abundance and Economic Growth," NBER Working Paper No. 5398 (Cambridge, Massachusetts: National Bureau of Economic Research).

Sala-i-Martin, Xavier, and Arvind Subramanian, 2003, "Addressing the Natural Resource Curse: An Illustration from Nigeria," IMF Working Paper 03/139 (Washington: International Monetary Fund).

Tseng, Wanda, and Robert Corker, 1991, *Financial Liberalization, Money Demand, and Monetary Policy in Asian Countries,* IMF Occasional Paper No. 84 (Washington: International Monetary Fund).

U.S. Department of Energy, 2002, *Oil and Gas Journal, Energy Information Administration.* Available via the Internet: http://www.eia.doe.gov/emeu/cabs/caspgrph.html.